SPIN-Farming Illustrated

By Wally Satzewich and Roxanne Christensen

This Is What Farming Looks Like!

ISBN: 978-0-615-58481-2

For information on licensing, foreign or domestic rights, or bulk copy orders for educational use, contact Roxanne Christensen at rchristensen@infocommercegroup.com, or at 610-505-9189.

Meet a SPIN farmer who's building a better future!

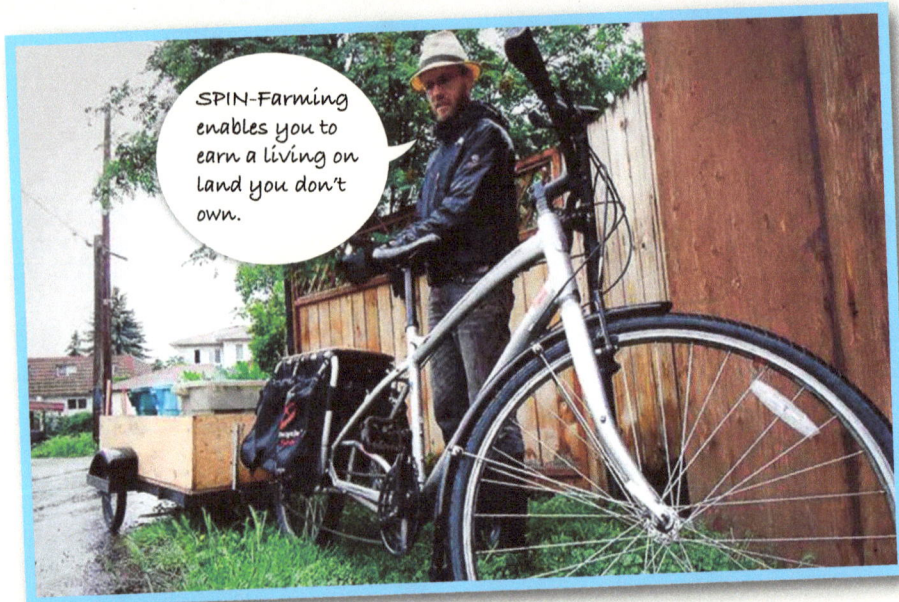

> SPIN-Farming enables you to earn a living on land you don't own.

Curtis Stone, Kelowna BC

A former band member of People for Audio, Curtis Stone is now the owner/operator of Green City Acres, a three quarter acre urban farm based out of Kelowna, BC. Having no prior experience as a farmer or gardener, Curtis started his farm simply out of a desire to be more autonomous, and to make a living by the triple bottom line principle. His farm employs one other person full time, and is still growing.

Curtis uses SPIN-Farming's multi-locational approach. His operation is spread out amongst eight different urban plots, rented from homeowners, to grow vegetables for local distribution. In exchange for the use of their land, landowners receive a weekly basket of produce throughout the growing season. This mutually beneficial arrangement saves them the burden and cost of maintaining a lawn, while reducing their food costs and providing Curtis with growing space. All of his produce is grown with natural methods, meaning that no chemical fertilizers, sprays, or pesticides are used. In fact, he barely uses any fossil fuels at all because his operation is entirely pedal powered. Bicycles and custom built trailers serve all his transportation needs, including fresh produce deliveries to restaurants and the Farmers Market, as well as the transport of compost and equipment, including a 400 lb. rototiller.

A bike tour down the west coast from Kelowna to San Diego in 2008 where Curtis visited off-grid homesteads, eco villages and urban farms, inspired him to try to make a difference. Using SPIN-Farming as a method of social change, Curtis is demonstrating that a transition away from a food system that is energy intensive, environmentally destructive and socially inequitable is not only easy and fun, but also profitable.

Green City Acres has now completed three successful seasons and Curtis is a case study example that the SPIN production system can be successfully implemented by anyone who wants to heed the calling to farm. During his off-season, Curtis works as a public speaker, teacher, and consultant, sharing his story and inspiring a new generation of farmers.

To see how others are bringing farming back home, visit *www.spinfarming.com*

Meet a SPIN farmer who's connecting with her past!

I don't have to leave my home. I'm here. I'm making money.

Brenda Sullivan, Glastonbury, CT

Like many of us, Brenda Sullivan grew up on farm stories, rather than a farm. She did not have to go too far back to find her farming heritage, though. Her father grew up on a farm, and he was the first generation to leave it. Her grandparents were the last to farm on her mother's side of the family. But all previous generations dating back to the 1500's were farmers on both sides. They were among the first families to go west in covered wagons looking for better land. But Brenda found her farmland right in her own backyard.

It was Brenda's daughter, Katie, who led her back to growing food. Katie had to be on the Ketogenic diet to control seizures. The children's hospital strongly recommended Brenda grow her own vegetables because food manufacturers could not be depended on to be honest about what they added to their products. Katie is no longer on that diet, but Brenda is grateful that it – and SPIN-Farming – helped launch her farming career.

After food gardening for 16 years in South Glastonbury CT, Brenda needed to supplement the family income due to the economic down turn. Because she cared for Katie, she needed a business that kept her close to home. So, using SPIN-Farming, she transformed her backyard into a 1,300 square foot farm, and she now sells her greens and specialty vegetables to restaurants and at two farmers' markets. She's a perfect role model for others like her who are turning to farming to re-skill, lower their grocery bills and plug the gaps that have been showing up in more and more household budgets.

Brenda now gives talks in support of local agriculture and is preparing workshops to help others like her get SPIN-style farms in and off the ground.

To see how others are bringing farming back home, visit *www.spinfarming.com*

Meet a SPIN farmer who thinks different!

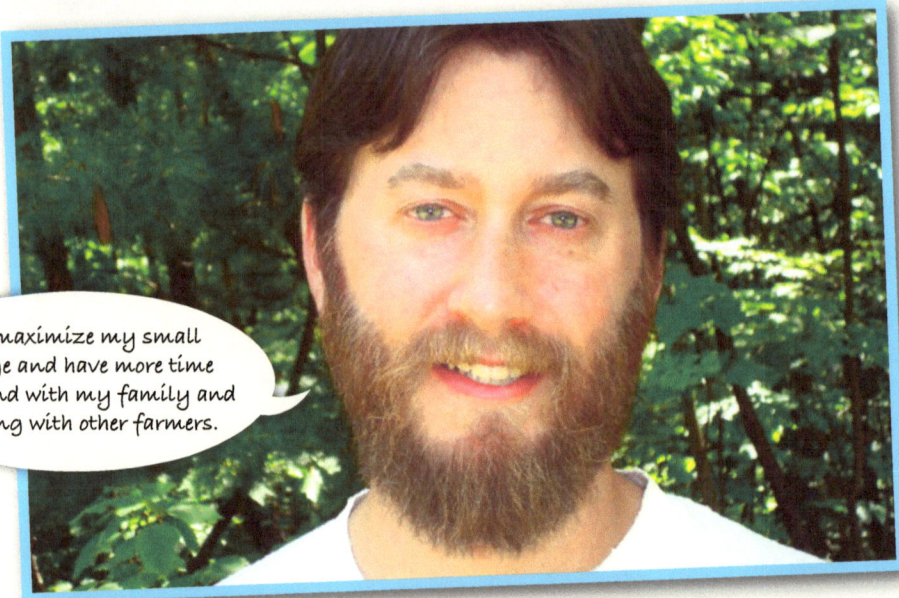

> I can maximize my small acreage and have more time to spend with my family and working with other farmers.

Andy Pressman, Jaffrey, NH

Raised in Columbus, OH, the home of many fast food chains, Andy Pressman has always searched for ways to grow healthier and sustainably-raised food for his family and community. Andy grew up with the mantra "bigger is better", the guiding principle of Midwest agriculture, but he always thought differently. He sought out farmers and teachers who would show him sustainable ways to build soil, grow food and make money. As a beginning farmer, he worked on diversified farms, raising vegetables, fruits, medicinal herbs and livestock. After a few years in the fields throughout the mid-Atlantic and New England, he returned to school to pursue his interest in Permaculture and small-scale agriculture, eventually earning a Master's degree in Sustainable Systems Design and Agroecology.

After school, Andy returned to farming in Virginia and then in Pennsylvania. In 2007, he became an Agriculture Specialist with the National Center for Appropriate Technology (NCAT), a non-profit organization whose mission is to promote and demonstrate small-scale, local and sustainable solutions to agriculture and energy. Moving to the city with a family and full-time career, Andy discovered SPIN-Farming, a system that would allow him to stay involved in farming by turning his yard into a commercial farm. So, as they say, when the student is ready, the teacher will appear. In the case of SPIN-Farming, the teacher was a computer, and SPIN's online learning series. But when Wally Satzewich, SPIN's developer, took to the road to do workshops, Andy was one of the SPIN master's first students. Shortly thereafter, Andy oversaw the implementation of the first University-based SPIN farm in the U.S.

Andy and his family now live in Southern New Hampshire, which is known as the Granite State because the soil is mostly made up of rocks, so farming small plots intensively is a big advantage there. He also continues his work with NCAT, conducting workshops and providing technical assistance on small-scale intensive farming. Andy says that because SPIN-Farming has taught him how to handle his time and land more efficiently, he makes more money, while having more time to spend with his family and working with other farmers.

To see how others are bringing farming back home, visit *www.spinfarming.com*

Meet a permie SPIN farmer!

> Permaculture needs wider adoption, and SPIN needs to develop ecologically. Together they can create lasting world change.

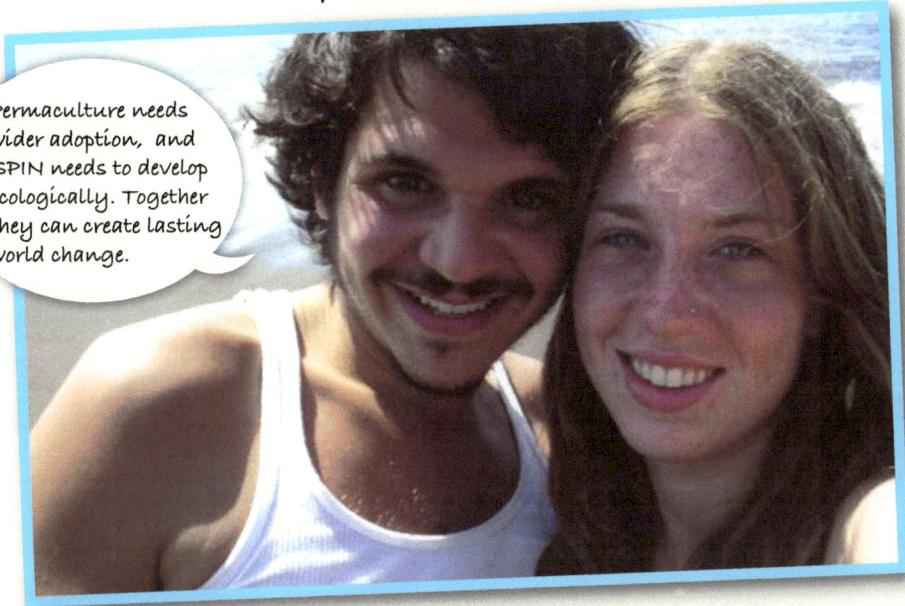

Shad Qudsi, Tzununa Guatemala

After graduating from Johns Hopkins University with a double major in Applied Math and Psychology, Shad Qudsi landed the corporate dream job but knew right away it was not for him. For Shad and his soon-to-be wife Colleen, following their dream led them to the north shore of Lake Atitlan in Guatemala. There they purchased a very small farm located in the traditional Mayan village of Tzununa and established Atitlan Organics, a SPIN and Permaculture-inspired, homestead that sells fresh fruits and vegetables to over 50 families, employs three full-time workers and maintains a profitable operating base all year round.

Shad is an avid permie. With Permaculture, he found a way to apply ideals he believed in to daily life. But what was missing was a way to make a living. There is only limited profit potential in teaching Permaculture, and what good is it to teach change at a two week retreat, only to send students back out to the status quo where they have to check their values at the door when they go to work? What attracted Shad to SPIN-Farming in addition to its low-input production model and efficient use of small land bases, was its business model.

At Atitlan Organics, Shad's income-producing SPIN plots are nestled very beautifully into his whole farm design which mainly focuses on greens, dairy goats and a large edible and useful plant nursery. Now he is sharing his experience and knowledge with others. If you'd like to learn SPIN-Farming in a place where the sun almost always shines, visit Shad at Atitlan Organics.

To see how others are bringing farming back home, visit *www.spinfarming.com*

This Is What Farming Looks Like!

In the pages ahead you'll be looking at the world through SPIN glasses. What you'll see is engaged, rather than escapist, farming. You'll see how crop production can be integrated into cities and towns, rather than segregated outside of them. You'll see innovative new land bases. You'll see how the same kind of intelligent, dedicated, craft-based farming, which has been kept alive for generations by a privileged few, can be practiced by anyone, anywhere.

You will see both what is possible, and what is practical. What is possible are communities that promote individual self-reliance while acknowledging the world's interdependence. What is practical is to make more from less.

You will see that SPIN farms are not the huge monochromatic blocks of golden wheat and tasseled corn that have come to define modern agriculture. Instead, a more intricate pattern weaves together sub-acre farmland. Individual plots of carrots, herbs, lettuce, radishes, spinach and tomatoes are framed by rows of sunflowers or fruit trees. Planting this carefully chosen diversity makes a pretty picture, and it also benefits the environment by eliminating the need for artificial pesticides and fertilizers and reconnects our lives to a more natural cycle.

These new urban and suburban farmsteads collectively define the kind of farming that anyone can understand and that more and more want to practice. So have a look at the world through SPIN glasses, and then go create your own farming vision, wherever you were planted, whether it's in the middle of an urban jungle, on the suburban fringe, or in a small town. Wherever you are, you can fit farming into your life. See how in the pages ahead »

SPIN-Farming® Illustrated

Section 1: Concepts 10

Section 2: Land Base 20

Section 3: Gear 30

Section 4: Work Flow 40

Section 5: Production 50

Section 6: Marketing 60

Lexicon

Looking Ahead

What's Your Vision?

SPIN-Farming® Basics

SPIN 2.0: Production Planning & Crop Profiles

About the Authors

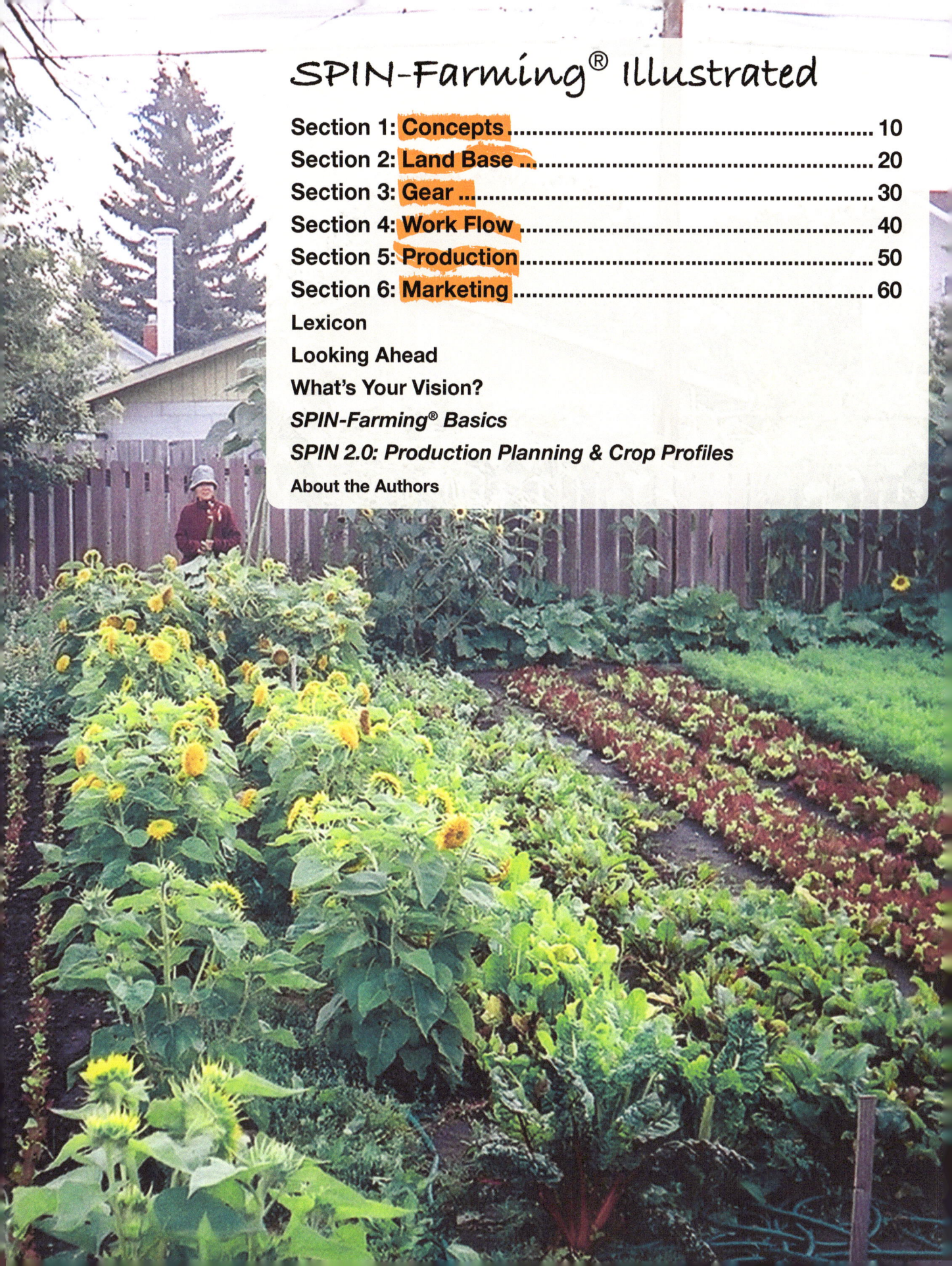

NEW FARMING ICON:

SUB - ACRE

PRODUCTION SYSTEM

D. Rose

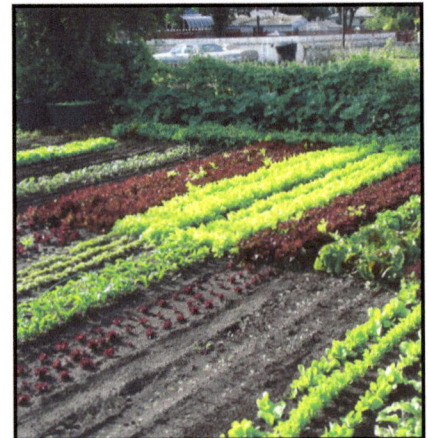

Section 1:

SPIN's Key Concepts

Use a systemized approach to planning and farm design.

There is an interesting tension, now, surrounding farming. Food is being extolled as the very essence of life at the same time as farmers are being put on the defensive to explain their methods. Rhetoric-laden food has replaced chemical-laden food.

To re-establish robust, locally-based, independently owned farms that can hold their own with industrialized agriculture, we have to get beyond discussion and debate and make farming attractive and viable again as a profession, even for those who live in cities and towns. SPIN-Farming is one way to do that.

SPIN-Farming greatly reduces the amount of land needed for commercial crop production.

Our culture currently defines a farmer as someone who uses expensive mechanized equipment and systems, employs large fleets of seasonal workers, and whose land base is hundreds, or even hundreds of thousands, of acres. SPIN (for s-mall p-lot in-tensive) offers an alternative, which is not in conflict with large scale agriculture, and is in fact symbiotic to it.

SPIN's approach to farming is a sub-acre one. The possibility of making a living on less than an acre of land certainly expands the options for new farmers, and SPIN-Farming provides a system that makes this possible. It is non-technical, easy-to-understand and inexpensive to implement and is now practiced widely throughout the US and Canada by those with no farming background and who may not have grown up ever connected to the land.

SPIN's growing techniques are not breakthrough. What is novel, however, is the way a SPIN farm is designed, implemented and run. Its system provides everything you'd expect from a good franchise: a business concept, marketing advice, financial benchmarks and a detailed day-to-day workflow. In standardizing the system and creating a reproducible process, it really isn't any different from McDonald's.

While most other farming systems focus primarily, if not exclusively, on agricultural practices, SPIN emphasizes the business aspects and provides a financial and management framework for having the business drive the agriculture, rather than the other way around. This keeps farmers focused on what matters most to their success. SPIN-Farming's key concepts include:

- Standard size bed – one that measures 2 feet wide by 25 feet long

- High-value crop – one that produces at least $100 gross per harvest per bed

- Intensive relay cropping – the sequential growing of at least 3 high-value crops in the same bed throughout the season

- 1-2-3 farm layout – the farm is divided into 3 different areas of cropping intensity

These concepts are contained in the SPIN-Farming lexicon. The sooner farmers become familiar with the terms and how they work together, the more productive and profitable they will be. The complete lexicon is included in the back of this book to help you start thinking and talking like a SPIN farmer.

Very small SPIN farms, in the 5,000 square foot range, devote most of their production to high-value crops in order to maximize their revenue. Larger farms, in the 20,000 sq. ft (1/2 acre) to 1 acre range, can include the production of both high-value and lower value crops. While its sub-acre scale makes SPIN-Farming particularly suitable for urban areas, it is not just an urban farming system. It is suitable for a wide range of contexts, and some SPIN farms combine urban, peri-urban and rural plots. The photo above is a typical urban backyard plot that measures 2,000 square feet. It is in intensive production, where 3 or more crops per year are grown. Containers of micro greens are in production at the back of the plot. This plot can produce close to $8,000 worth of vegetables per season.

The plot at right is a rurally-based single crop area of onions and scallion. It, too, uses standard size beds. The 2 foot width is easily straddled with the legs to make quick work of planting, weeding and harvesting, and it also accommodates the width of many rototillers. The 25 foot length corresponds to many types of garden sprinklers. The bed unit of production is tied to SPIN's revenue targeting formula which SPIN farmers use to plan and produce a steady and significant cashflow throughout the season.

Small Plot = Big Money

SPIN's revenue targeting formula:

1 acre = 480 standard size beds

Intensive relay production = 3 high-value crops harvested/bed/season

@ $100 value/harvest each bed = $300 gross revenue/season

480 beds x $300/bed = $144,000 potential gross revenue/acre/season

NEW FARMING ICON:

NICK'S

MULTI-LOCATIONAL

FARM

DAVE'S

ANDREA'S

D. Rose

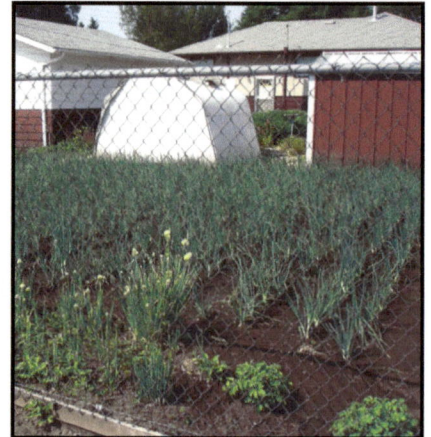

Section 2:
SPIN Land Base

You don't need to own much, if any, land to get started in farming.

This plot is 50 feet x 100 feet. It is an urban backyard and shares a fence line with a neighboring property. It had been an abandoned lot and was purchased by the farmers for $1,000. Half of it will be used for soil-based production for about 2.5 SPIN segments, which will be put to long season crops such as winter squash, pumpkin, and tomatoes. The other half will be container-based. Revenue potential is $3,000 - $5,000 per season. Not a bad return on investment.

What do you see when you put on your SPIN glasses?

Food production everywhere!

SPIN-Farming offers a new approach to land access, which is one of the biggest barriers to entry for new farmers. Conventional thinking is that you need to purchase land in a rural area. But once you put on your SPIN glasses, you start seeing, and thinking, differently. How about renting or bartering backyard garden plots that you can use as your land base? How about converting your lawn into a front yard farm? How about a single-sited suburban farm? Roof tops and parking lots are starting to sprout urban farms. The point is that not owning land, or city living, do not have to prevent you from becoming a farmer. In fact, they give you big competitive advantages like micro climate, availability of existing water sources and close proximity to lucrative markets.

The backyard of your home is a good place to start a multi-locational farm. Everything is in play here – small plots, vertical farming, container growing, greenhouse production and composting.

SPIN-Farming is practiced in a wide variety of contexts – from larger acreages in the country to raised beds on sub-acre lots in the city. Every situation has its advantages and disadvantages. The point is to accentuate the positive. For instance, a multi-locational farm spread out over many plots, poses some logistical considerations, but has these advantages: each plot has its own micro climate so that you can grow a wide variety of crops; you can manage crop rotation more easily and effectively, and it offers a form of risk management since any problems are limited to specific plots. Yardsharing for the purpose of commercial farming is becoming more common, and some SPIN farmers report they have more land offers than they can handle.

Another point to understand is that if a novice farmer cannot master production on a sub-acre plot, their chances of success will not be increased by having even more land and overhead to manage. The advantage to SPIN that many miss is that it can increase the survival rate of new farmers who might actually be undermined by buying into the old model. Those who give up might otherwise have succeeded if they weren't initially overburdened financially by debt and operationally by large acreage and overhead.

While urban sprawl continues to swallow up more tracts of prime farmland, there is a counter balancing process at work. Urban and suburban farms create new cropland amid concrete. Much of the land that is suitable for agriculture within cities and towns is relatively small in size and scattered. Conventional farming can't take advantage of these spaces, but SPIN-Farming can. Its well-thought out system can be used to reclaim these spaces for commercial agricultural production.

Establishing any new idea requires a leap of faith, but with SPIN-Farming, advocates for local food systems now have proof of concept. SPIN-Farming has been around for 6+ years, and many people are having success with it, so it provides a track record that can be replicated. It is also helping to professionalize an activity that historically has been fragmented, unorganized and unrecognized, but which is now part of a $1.2 billion industry in the U.S. SPIN's land base may be garden-size, but its net income is the same, or greater than, large scale conventional farmers, but with a lot less stress, and with a lot more certainty of success from year to year.

Urban/Suburban Plots = Prime Farmland

How SPIN-Farming Can Help Revitalize Big Cities and Small Towns

- It makes fresh, healthy food accessible to consumers.
- It makes use of under-utilized land such as back and front yards, and vacant side lots.
- It beautifies blighted areas.
- It generates economic activity.
- It helps close the loop on compostable waste.
- It lowers the carbon foot print of food production.
- It reconnects people to the natural environment.

In the end, everyone benefits - communities, consumers and farmers.

New Farming Icons:

Rototiller

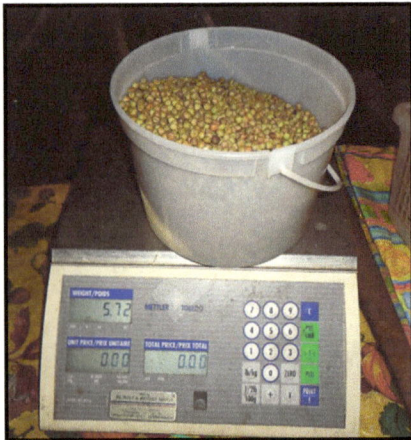

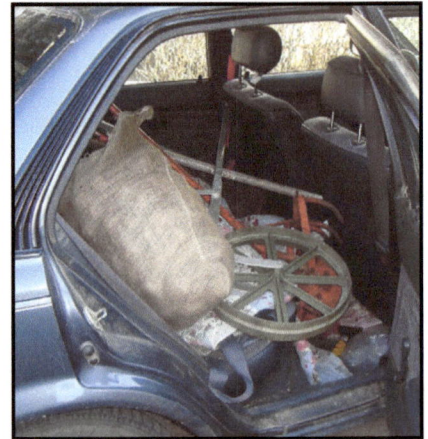

Section 3:

SPIN Gear

You don't need to take a big financial risk to get started in farming.

The only mechanized equipment backyard, front lawn and neighborhood lot farmers need is a rototiller. A tiller is both easy on the pocketbook and the environment. It is portable, highly maneuverable and easily maintained. Its tank can be filled with a gallon of gas, which generates hours of productivity.

SPIN's
bottom line:
Little or
no debt.

Conventional farming requires a substantial amount of investment. Tractors can cost tens of thousands of dollars, or even hundreds of thousands. And that is just the start. Combines, harvesters, irrigation systems requiring central pivots and wheel lines, all represent an investment of the mid- to-high six figures.

SPIN-Farming, however, greatly reduces the amount of capital needed to start a farming operation. Top-of-the-line walk-behind tractors or rototillers cost around $5,000. Commercial coolers are also a low four figure purchase. Investing $10,000 in your SPIN-Farming business will get you a long way. $10,000 gets you nowhere in conventional farming. SPIN farmers also have a wide range of options available when choosing how much or how little to spend on any one investment. Used, recycled and do-it-yourself items are their stock in trade.

Harvesting on sub-acre plots is very low tech. Scissors or a sharp knife and rubber bins are all that are needed, along with strong hamstrings. Fitness is a perk of the job.

This is a work station for the post-harvest handling of garlic in an urban backyard. Tables for drying and braiding are disassembled when the work session is over.

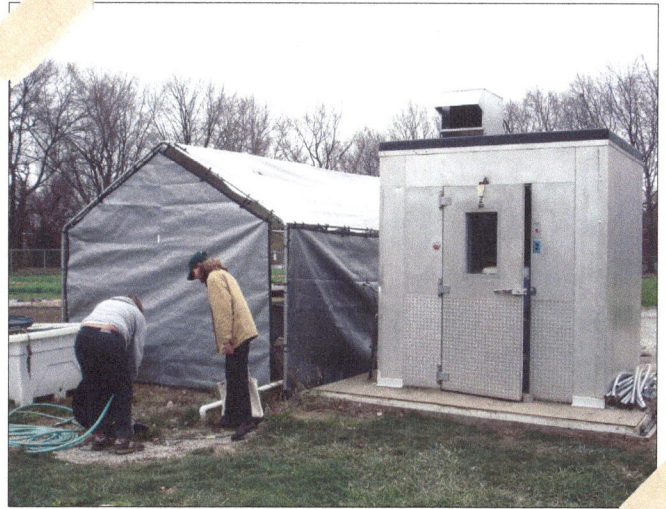

Above are two main components of a SPIN farm's infrastructure – a post-harvesting work station and cooler. SPIN farmers who take the high-road invest in a commercial grade cooler which is crucial for high-quality crop production and keeping control of work flow. The optimal size for a commercial cooler is about 10'x 10', and the cost including installation should be around $5,000.

A less expensive option is an upright produce cooler. A used one can be purchased for under $1,000. It can be plugged into a standard wall outlet, and it is modular – you can purchase additional ones as your operation expands. There are also DIY cooling systems.

Work stations, too, do not need to be expensive or even permanent. This work station is a tarp spread over some poles, and it cost around $1,200. An important consideration is access to potable water. In an urban setting this will usually not be an issue, and this is one of the great advantages of urban farming. Good work station hygiene is important, regardless of whether the work station is simple or elaborate. Sanitizing bins with proper cleaners is a must, as well as regular cleaning of work areas.

This peri-urban 7,000 square foot site is a 5 minute drive from a city of 200,000. Fencing can be a considerable expense. There are many options, including electric ones. This one was homemade by the plot owner, who is a metal worker. His shop is adjacent to this site. The fence is effective for deer, but small critters, like gophers, can pose big problems. Specialized production of onions, potatoes, and squash addresses this challenge since the gophers at this location have not acquired a taste for these crops (yet). When they do, a different crop will be planted.

In urban areas, two legged critters are the main threat. Always looking for ways to turn minuses into pluses, city farmers turn fence lines into profit centers by using them for the production of vertical crops. Some, like Chartenais melons, are quite exotic, at least to the urban consumer, and that translates into high demand and pricing.

Some urban farmers employ decoy crops. These are crops for the taking that are grown around the permimeter of their farms. They aren't calculated into production.

SPIN farmers equip themselves with many of the same tools that gardeners use. But specialty hoes, like a stirrup hoe and a colinear hoe, are not found in the typical garden shed. They save lots of time (and backs) when it comes to weeding. The cost of a full line of SPIN hand tools is around $1,000. Most are one-time purchases and will outlast many farmers.

NEW FARM ICON:

STOP WATCH

D. Rose

Section 4:

SPIN Work Flow

SPIN's systematic work flow makes the wide variety of farming tasks manageable, satisfying and relaxing.

Work flow is one of the most important aspects of any farming operation, big or small. Because SPIN farms are owner-operated and aim to minimize or eliminate the need for outside labor, work has to be well-managed for the farm to be sustainable from a logistical perspective. Here, sustainable means work flows that you can handle without burning yourself out half way through the growing season, or after a few years.

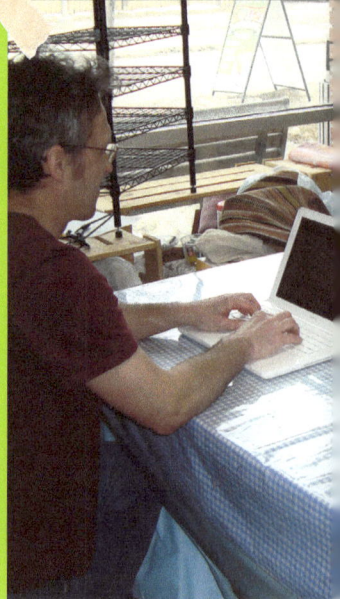

Commercial refrigeration allows produce to be harvested throughout the week. Let's say you are attending a Saturday farmers market, and are targeting $1,500 in sales. This requires a substantial amount of produce. Without refrigeration capacity, you would start harvesting and prepping late in the week with most of the work occurring on Friday, and possibly some very early the day of market. This makes for an extremely compressed, high-pressure work flow. Production volume is sacrificed because you only have so many hours in a day. It could be boosted by hiring labor, but that is an extra expense and can be a management headache. With SPIN's high-road approach, you can spread harvesting over a five day work week. Carrots, scallion, potatoes and other crops can be harvested early in the week, Monday through Wednesday, while more perishable crops, such as salad greens, can be harvested, prepped and bagged on Thursday and Friday. Ideally you should be done your harvesting and prepping by Friday afternoon and curl up on the sofa Friday night, just like any other working stiff.

Spreading out your harvesting over many days means that you won't have an excessively long harvest and prepping session on any given day. This gives you time to perform other tasks such as weeding, watering, pest control and replanting. Your days are balanced among a variety of activities. SPIN farmers are never bored.

Since salad greens are one of the main crops of many SPIN farms and among the most labor intensive, establishing work rates for this crop is important.

A concept that ties closely to work flow is work rate. To plan your work week, you need to determine how long it takes you to do a variety of tasks, from harvesting to watering to weeding and planting. That is your work rate. For instance, how long does it take to harvest a standard bed of spinach? How long does it take to wash 20 lbs. of spinach? How long does it take to bag it into 1/4 lb. bags? Knowing your work rate to harvest one bed will allow you to project how long it will take to harvest 10 beds. This type of knowledge allows you to schedule your work sessions throughout the week.

Knowing work rates also allows you to set benchmarks and helps you decide if and when to use outside labor. If the people helping you are not achieving work rate benchmarks, then you will know that it is not worth the expense to have them help you.

Work rate analysis can be taken to a higher level by determining how long it takes to accomplish tasks with one, two, or even three people working together, and whether multi-person teams can get more accomplished per person than one person working alone. In this case the tasks involved are broken down into discrete steps.

What is the thinking that goes into planning a typical work session? Let's use one of SPIN's most important crops as an example – spinach. You have 40 lbs. of spinach in the cooler, and you set up a work session to get the spinach bagged. Such a session might take 1.5 hours with one person doing the bagging and putting on the twist ties. But you might find that putting on the twist ties takes as long as putting the spinach in the bag. So if you specialize the tasks, with one person putting the spinach in a bag, and then another person putting the tie on the bag, you might discover that when two people work together for this type of work session, the 40 lbs. of spinach might get bagged in half the time it would take for one person.

So work session planning involves determining when specializing certain tasks is worth the extra manpower involved. That requires establishing one person/two people work rates for certain types of tasks. A detailed understanding of work rate will allow you to use outside help strategically, so that you know where it is best used, so that you get the most bang for your buck.

SPIN is for the Whole Family!

SPIN very much preserves the farming tradition of tapping into the low cost or no cost network of family and friends during peak labor times. Keep in mind that older people have a better work rate than younger people for certain tasks. Work that can be done from a seated position and involves focus and patience, like cleaning and bagging, is often performed better by older people.

NEW FARMING ICON:

STRADDLE

D. Rose

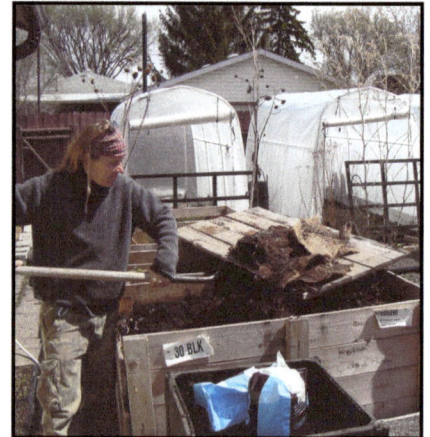

Section 5:

SPIN Production

The SPIN-Farming system is based on a standard size bed, which measures 2' wide by 25' long.

Here you see the standard size bed's advantage. It can be easily straddled with the legs which makes quick work of planting, weeding and harvesting. For practical purposes, bed width is based on the width of your tiller unit. If your tiller width is 20 inches, then make your beds that width.

The point is to know your site and to be flexible in your thinking on how to best utilize it.

SPIN-Farming is based on the bed system. Beds are made with a reartine tiller, and no special effort is made to raise the bed. Many new farmers mistakenly assume that raised beds are the answer in all situations. But raised beds can dry out quickly in arid climates, and they cost time and money to build. There is no inherent production advantage to them. Careful attention must be given to bed width, length, and walkway width. Frequently, different bed widths and lengths will be used on the same farm. You might have an area that is prone to flooding and decide to use raised beds in that area. A multi-locational farm might have different bed configurations at each of its sites. Beds wider or longer than the standard size bed can be appropriate for certain contexts, such as areas that are soilless or on rooftops. It is important to note that just because a 4' wide bed is used in certain areas does not mean that SPIN-Farming is not being applied.

This is a wide bed of radish production that measures about 3' wide. In SPIN there are many possible types of bed configurations. The more experienced you become, the more variations will occur to you. SPIN farmers are highly ingenious and are always coming up with new and inventive ways of improving the system.

Tilling is necessary in commercial systems. Land has to be put into production quickly, without spending inordinate amounts of time preparing beds with hand tools. Shallow tilling is necessary for many types of leafy greens. A shallow tilling might only require one or two passes with the tiller. Deeper tillings, which are required for carrots and garlic, might require three or four passes, depending on the soil conditions.

If overtilling is a concern, hand tools can be used instead. Soil can be spaded over, and then raked flat. Hand cultivators can be used to loosen up the soil. In intensive relay areas where there is a lot of cropping, beds can initially be made with a rototiller, and subsequent bed preparation can be done using hand methods. No till areas are also possible with the use of a transplant-based system.

Beds can be permanent, semi-permanent, or non-permanent. Permanent beds are areas that maintain their configuration from season to season. Semi-permanent areas change their configuration from year to year. Non-permanent areas change their configuration several times during the year. For instance, you might have a segment planted to spinach using one bed/walkway setup, and then change the setup for cauliflower, which uses wider walkways. Once you establish work rates on bed prep, you will be surprised at how quickly an area can be reconfigured.

SPIN-Farming is not based on any one approach or formula for soil fertility because so much depends on local conditions. You should consult with local growers or agriculture experts to develop a fertility program. Soil tests are always a good idea because in addition to identifying what amendments are needed for fertility, they can also address soil contamination. In urban areas, heavy metals are a concern, especially in older cities with an industrial history. Lead paint from demolished buildings can also be an issue.

Soil can be kept healthy by using inexpensive, locally available organic soil amendments. Many organic soil amendments are sold in 50 lb. bags as animal feeds and can be purchased at local farm supply stores. Fertilizers like dried molasses and oil seed and blood meals break down quickly in the soil and are relatively quick acting. Others, like alfalfa pellets and sugar beet pulp, are good soil conditioners and builders as well. They are easy to apply, and because of their concentrated nature, they are not required in large quantities. Other fertilizer options are fish meal products, compost teas, specialty foliar sprays and your own compost.

SPIN-Farming is based on organic methods of pest control, but as with soil prep and fertility, it does not dictate specific techniques. Each SPIN farmer is encouraged to experiment to determine which are most appropriate for their scale production, and which fits best into their context. Using a backpack sprayer with rotenone does not work well in an urban scenario, but it usually works well in a rural one. You need to be sensitive to your immediate surroundings and act appropriately and considerately. Keep in mind that pests refer to insects, as well as larger members of the animal kingdom, everything from gophers to deer. At a sub-acre scale, farmers are constantly working and monitoring their plots so any pest problems can be idenitfied and dealt with easily and quickly, such as by hand picking, shown above.

The sub-acre advantage also comes into play with irrigation. Unlike large-scale farms which have to rely on one system, such as a central pivot or wheel line which cost many tens of thousands of dollars, SPIN-Farming relies on simple home gardening equipment that is used in innovative ways. Because SPIN's relay cropping involves a lot of diversity with crops at all different stages of growth, an irrigation system must be flexible. You might be surprised at how far you can go with simple hand watering. Yes, it is labor intensive, but hand watering is very efficient in terms of water usage and of placing a direct flow of water to exactly where it is needed. The guiding principles are that SPIN irrigation systems should be practical, appropriate and cost-effective.

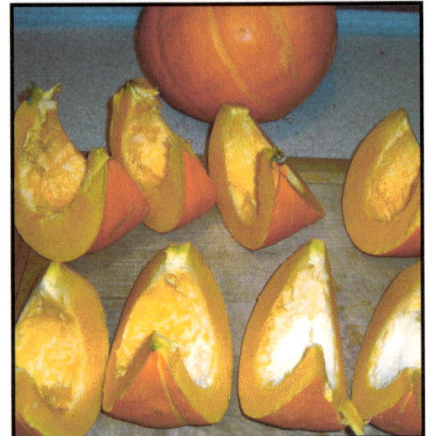

Section 6:

SPIN Marketing

They have yet to breed a vegetable that sells itself.

This is an indoor stand with single tiered pricing. All items are the same price here, $3.00 per item or any 2/$5.00. Since this requires pre-bagging, you need to include bagging work sessions into your weekly schedule.

SPIN farmers sell direct to their customers, and the way they get top prices for their crops is to give them "eye appeal."

Getting good crop prices is more in your control than you think. What it takes is creative thinking. Never assume that pricing is set in stone. Fixed pricing is more common to larger scales of production. As a farmer selling directly to the public, you have the advantage of being able to vary your pricing according to the volume of your production, and your needs. One example of innovative pricing is SPIN's mix-and-match multiple unit pricing which offers a variety of produce items at the same price tier. The idea is to get the customer to spend more money at your stand, as well as making shopping an easier experience for them. An example is $3.00 per item, or any 2 for $5.00. With this type of unit pricing, a customer knows exactly what they are going to get for $5.00, and they look forward to considering their choices.

Multiple tier pricing is another effective technique. This is mix-and-match pricing at two different price tiers. In one case you might have any 3 items for $4.00, and on a separate table have items priced at any 3 for $10.00.

Here is a typical purchase. 2 pumpkin slices, and 4 1/4 lb. bags of salad greens - $15 total. A pricing system makes it easy for customers to spend their money. It also means quicker checkout time. Loose produce makes for slow processing time and frustrated customers.

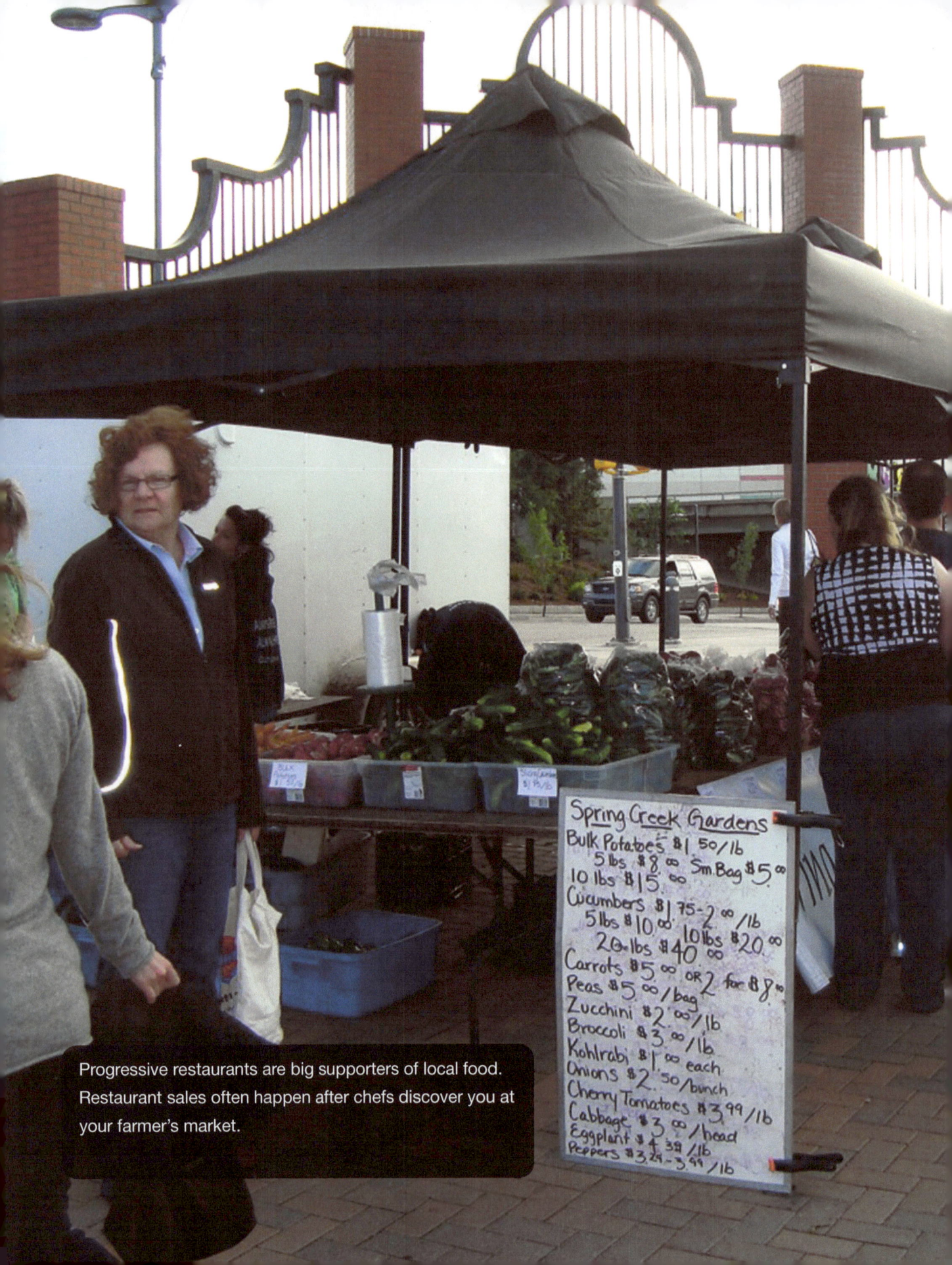

Progressive restaurants are big supporters of local food. Restaurant sales often happen after chefs discover you at your farmer's market.

Spring Creek Gardens
Bulk Potatoes $1.50/lb
5 lbs $8.00   Sm. Bag $5.00
10 lbs $15.00
Cucumbers $1.75-2.00/lb
5 lbs $10.00   10 lbs $20.00
20 lbs $40.00
Carrots $5.00 OR 2 for $8.00
Peas $5.00/bag
Zucchini $2.00/lb
Broccoli $3.00/lb
Kohlrabi $1.00 each
Onions $2.50/bunch
Cherry Tomatoes $3.99/lb
Cabbage $3.00/head
Eggplant $4.39/lb
Peppers $3.29-3.99/lb

Much of SPIN-Farming's marketing techniques are based on access to well-attended farmers markets. But new farmers may not have access to farmers markets so other options need to be created.

Home deliveries can work for new growers, but they require some investment in creating a distribution system. Email marketing, whereby potential customers are solicited by email and then sent a weekly list of available produce which is picked up at a central point or delivered at an extra charge, can also be used. A Community Supported Agriculture (CSA) program, where a share of the harvest is sold in advance of the growing season, is an option, but these require steady, consistent production which is only obtained after several years of experience. Some novice growers gain initial CSA experience with friends and family who are willing to regard their share as a learning experience for the farmer. Restaurants are another option, but they, too, require the ability to produce steadily and consistently. When it comes to marketing channels, you need to research the options that are available in your area, and decide which are appropriate to your level of experience, and be prepared to develop your own.

SPIN-Farming provides a business model that generates steady consistent revenue for a defined period of time. A revenue target is decided and then strategies are put in place to achieve that target based on marketing weeks. The more marketing weeks you have, the easier it will be to achieve targeted revenue. A typical range for marketing weeks is 20 – 30. Early spring production, as seen in the plot above, can generate thousands of dollars in a few good marketing weeks. Conventional approaches do not consider this period a significant contributor to revenue. But early spring is when supplies are limited, which supports high prices.

SPIN-Farming offers revenue benchmarks based on experience level. Novice SPIN farmers, those who are in years 1-3, target $500 gross during most of their marketing weeks. Achieving or even exceeding $15,000 your first year is a significant accomplishment. As your experience increases, so does your revenue. Expert SPIN farmers can achieve $2,000+ gross per week.

For more information on SPIN-FARMING please visit *www.spinfarming.com*

Home/Community-based Business + Farming = Great Opportunity

SPIN farmers think of themselves as business people who happen to be growers. What they are doing is re-casting farming as a small business in a city or town. As SPIN-Farming becomes more commonplace, it will start to again be obvious where real food comes from and why it is better. This will expand and solidify the already rapidly developing markets that will sustain local food systems long term.

What is ironic about a culture that invents a word like "locavore" is the complex effort it takes to achieve something that was once straightforward, and the amount of hype that surrounds what was once unspoken. For SPIN farmers, the simple words "trust me" are all that's needed.

SPIN FARMING®
Lexicon

SPIN-Farming has its own unique techniques and language. To help get your head around how SPIN differs from conventional farming methods, or from home gardening, here's a translation of the important terms you'll be using as a SPIN Corps member.

Sub-acre land base – SPIN transfers commercial farming techniques to sub-acre land bases. Farmers do not need to own much, or any land, to start their operations, and they can be single or multi-sited.

Structured work flow practices – SPIN outlines a deliberate and disciplined day-by-day work routine so that the wide variety of farm tasks can be easily managed without any one task becoming overwhelming.

High-road/Low-road – SPIN distinguishes between two different harvesting techniques. High-road utilizes commercial refrigeration equipment. Low-road harvesting does not.

High-value crops – SPIN devotes most of its land base to the production of high-value crops, defined as ones that generate at least $100 per crop/per bed.

Relay cropping – SPIN calls for the sequential growing of crops in a single bed throughout a single season.

Intensive relays – 3 or 4 crops per season are grown.

Bi-relays – 2 crops per bed per season are grown.

Single crops – 1 crop per bed per season is grown.

1-2-3 bed layout – Refers to the 3 different areas of a SPIN farm devoted to the different levels of production intensity.

75/25 land allocation – Dictates how much land is assigned to the different levels of production on a SPIN farm. The aim is to balance production between high-value and low-value crops to produce a steady revenue stream and to target revenue based on farm size. The smaller the land base the more of it can be devoted to intensive relay production.

Farm layout – SPIN provides guidelines for segmenting a land base into a series of beds, separated by access alleys, which are small 2 feet strips, just wide enough for a rototiller. An acre accommodates approximately 480 standard size beds, including the necessary paths and access alleys. SPIN can also incorporate more traditional approaches to land allocation.

Standard size beds – SPIN utilizes beds measuring 2 feet wide by 25 feet long.

Revenue targeting formula – By growing high-value crops worth $100 per harvest per bed, and by practicing intensive relay cropping which produces at least 3 crops per bed per season, SPIN targets at least $300 in gross sales per bed per season. With approximately 480 beds per acre, the maximum revenue potential is 480 beds x $300 per bed per season = $144,000 gross sales per acre. When farming is approached in terms of beds instead of acres, the result is a very precise idea of how much growing space can be utilized, and how that space can be managed to generate predictable and steady income.

Organic-based – SPIN relies on all-organic farming practices. There are minimal off-farm inputs and very little waste.

Crop Diversity – A SPIN product line contains a very wide diversity of crops, with some SPIN farms producing over 100 different varieties and 50 different types of crops per season. However, SPIN also provides models that specialize in a particular crop.

Season extension is optional – SPIN does not rely on season extension to expand production; however season extension can be utilized to push SPIN yields and income significantly higher.

Direct marketing – SPIN bases crop selection on what local markets want. Being close to markets allows for constant product feedback and ensures a loyal and dependable customer base. Grow what you sell, don't sell what you grow, is the SPIN farmer's mantra.

Mix and match multiple unit pricing – SPIN's marketing approach is to pre-bag produce items and sell them at certain price tiers – for example, $3.00/unit or any 2 for $5.00.

Commercial refrigeration capacity – SPIN calls for taking the "high-road" by utilizing commercial refrigeration capacity because cooling crops immediately after they are harvested retains their quality which supports premium pricing. It also provides control over the harvest schedule and allows for a manageable work flow.

Minimal mechanization and infrastructure – SPIN's most important and costly equipment are a rototiller and a walk-in cooler or upright produce cooler. All other SPIN implements and infrastructure can be sourced at local garden supply or hardware stores.

"Home-based" work crew – Supplemental labor requirements for a SPIN farm are minimal and can be readily obtained within the network of family, friends, or within the local community.

Utilization of existing water sources – SPIN relies on local water service or wells for all of its irrigation needs.

Low capital intensive – Minimal infrastructure and minimal overhead keeps SPIN farm's start-up and operating expenses manageable. The bottom line is little or no debt.

Looking Ahead...

Family farmsteads passed down from generation to generation. A lone tractor silhouetted against the horizon. Rolling hills of corn and grain. These classic images have defined farming for over a century.

But now, for reasons both pragmatic and profound, farming has become a catalyst for inventive activity by entrepreneurs, technologists, artisans, change agents and the neighbor next door. In their hands, it is being broken apart and rebuilt.

SPIN-Farming is one representation of this. There are, and will be, many more. Whatever your farming vision may be, go create the future icons that this time and place will be known by.

A New Way to Learn to Farm

The agricultural system we now have was built on the fact that fewer people each year wanted to farm. This has been the reality in the U.S. for several generations, and it has led to the capital and chemical intensive agriculture and centralized supply chains that produce more food with fewer people. A system that counterbalances industrial agriculture will require millions of more farmers. Many claim that will never happen. But look around...

Realize what is possible...

Do what is practical...

More and more people are heeding the call to farm. They don't come from traditional farm families. They don't own much – or any – land. Some have been educated in other professions, or have had other careers. Some have home or community gardening experience, but others have never had dirt under their fingernails. Some farm in their backyards in the city. Others do it on front lawns in the suburbs. Some do it part-time; others full-time. Some have more money than they know what to do with, and others have less than they need. Some are convinced the world is doomed, and others are trying to save it. Some are trying to recreate the past, while others are looking to build the future.

What unites them all is a vision of meeting more food needs locally. That calls for a new vision of farming. Whatever your vision happens to be, have a look at these other publications from SPIN-Farming, and then start farming, wherever you were planted »

What's possible...

What's practical...

Basics

Get with the system!

SPIN-Farming® Basics has already helped hundreds of new farmers get started in business. While each implements SPIN differently, they are all pioneering a new way to farm, one which everyone can understand and more and more want to support. They share some of their stories at the SPN-Farming website and also generously offer advice and guidance in SPIN's online support group. When you start practicing SPIN you are welcome to join them. But first, you need to get with the system.

Here are some of its benefits:

- It greatly reduces farm startup and development time.
- It eliminates much trial and error.
- It shortens the learning curve.
- It provides specific benchmarks to measure progress.
- It provides more control over outcomes and income.
- It makes collaboration easier.

While other farming books address growing, marketing and management, *SPIN-Farming® Basics* ties them all together in a framework that keeps you focused on what matters most to making your business a success.

SPIN-Farming® Basics
$83.93

Available for immediate download at *www.spinfarming.com* or order a print copy through your favorite bookstore

...garlic

...radish

...cipollini onions

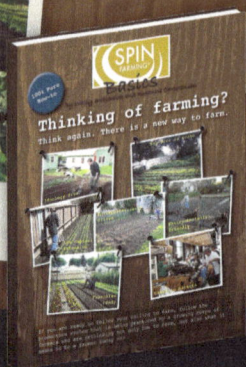

JOIN IN!

A society not only must have its ideals; they must from time to time be re-constituted in a way that sparks our imaginations and guides our strivings. That is what is happening now. Across several countries, at countless sustainability forums, eco-expos and green gabfests, people are coming together to think things through again, make certain re-appropriations, create a new ideal. For many, especially those in cities and towns, this new ideal includes rebuilding local food systems. While the last few decades have seen many last harvests as cropland was turned into concrete, SPIN is for those who tend cropland amid concrete. May *SPIN-Farming® Basics* pave the way to your first harvest.

Production Planning & Crop Profiles

Scale Up Your Production and Income!

Once you have mastered the basics of SPIN-Farming, , you can scale up your production and target higher levels of income. SPIN-Farming 2.0 provides the advanced sub-acre concepts, financial benchmarks and data to do that. Learn how to plan for expanded production of a wide array of crops. Use first-ever benchmarks to set goals. Plan production and target revenue with data on 40 classic SPIN crops including:

- days to harvest
- bed/walkway configurations
- in-bed spacing
- number of plants per row, per bed and per segment
- yield per row, bed and segment
- average seed count per pound
- seed required per row, bed and segment
- cost to seed a bed and segment
- crop value status
- price tier suggestions
- targeted revenue per bed and segment

Building on the non-technical, easy-to-understand and inexpensive to implement farming system outlined in *SPIN-Farming Basics*, *SPIN 2.0: Production Planning & Crop Profiles* provides the detailed data and analysis needed by the new breed of farmer whose success depends on producing good food at a reasonable profit.

STREET VALUE FARMING

Potatoes...$1,300

Onions...$2,431

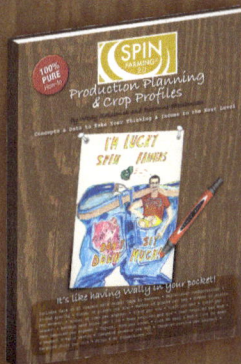

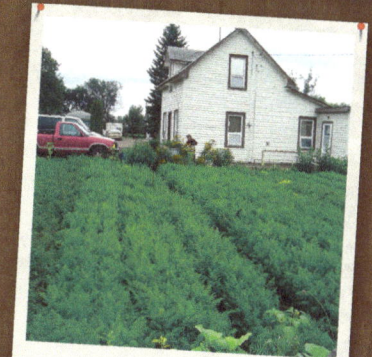

Carrots...$2,437

SPIN 2.0: Production Planning & Crop Profiles
$69.98

Available for immediate download at *www.spinfarming.com* or order a print copy through your favorite bookstore

About the Authors

Wally Satzewich

Wally Satzewich and Gail Vandersteen operate Wally's Urban Market Garden which is a multi-locational sub-acre urban farm. It was originally dispersed over 25 residential backyard garden plots in Saskatoon, Saskatchewan, that were rented from homeowners. The sites range in size from 500 square feet to 3,000 square feet, and the growing area totals a half-acre. The produce is sold at The Saskatoon Farmers Market and restaurants in the city.

Wally and Gail initially started farming on an acre-size plot outside of Saskatoon 20 years ago. Thinking that expanding acreage was critical to their success, they bought some farmland adjacent to the South Saskatchewan river 40 miles north of Saskatoon where they eventually grew vegetables on about 20 acres of irrigated land. The farmland was considered an idyllic farming site on its riverfront location. However, the crops were perpetually challenged by wind and hail, insect infestation, rodents and deer. Fluctuating water levels inhibited irrigation during dry spells. "We still lived in the city where we had a couple of small plots to grow crops like radishes, green onion and salad mix, which were our most profitable crops. We could grow three crops a year on the same site, pick and process on-site and put the produce into our cooler so it would be fresh for the market," Gail says.

After six years farming their rural site, the couple realized there was more money to be made growing multiple crops intensively in the city, so they sold the farm and became urban growers. Growing vegetable crops in the city was less complicated than mechanized, large-scale farming. They used to have a tractor to hill potatoes and cultivate, but they discovered it's more efficient to do things by hand. Other than a rototiller, all they need is a push-type seeder and a few hand tools.

They have recently expanded their multi-locational vegetable and flower gardens in the hamlet of Pleasantdale,Saskatchewan which will serve as the home base for training programs on sub-acre farming.

Wally points out that urban growing provides a more controlled environment, with fewer pests, better wind protection and a longer growing season. "We are producing 10-15 different crops and sell thousands of bunches of radishes and green onions and thousands of bags of salad greens and carrots each season. Our volumes are low compared to conventional farming, but we sell high-quality organic products at very high-end prices." The SPIN-Farming method is based on Wally's successful experiment in downsizing and emphasizes minimal mechanization and maximum fiscal discipline and planning.

Brian Halweil, a food issues writer and researcher at the Washington-DC-based Worldwatch Institute, interviewed Wally and referenced his farming approach in Eat Here, which documents worldwide initiatives in building locally-based food industries.

About the Authors

Roxanne Christensen

Roxanne Christensen co-founded Somerton Tanks Farm, a half-acre demonstration urban farm that served as the U.S. test bed for the SPIN-Farming method from 2003 to 2006. The farm, which was operated in partnership with the Philadelphia Water Department, received the support of the Pennsylvania Dept. Of Agriculture, the Philadelphia Workforce Development Corp., the City Commerce Department, the USDA Natural Resources Conservation Service, the Pennsylvania Department of Environmental Protection, and the Pennsylvania Department of Community and Economic Development.

In 2003, its first year of operation, the farm, located in the sixth largest city in the U.S, produced $26,000 in gross sales from 20,000 square feet of growing space. In 2006 gross sales reached $68,000. In just four years of operation this demonstration farm achieved levels of productivity and financial success that many agricultural professionals claimed were impossible.

Based on the agricultural and financial breakthroughs that were demonstrated at Somerton Tanks Farm, the state of Pennsylvania funded an economic feasibility study that documented the urban farm's economics and projected its maximum income potential to be $120,000 from under an acre of growing space.

As co-author of the SPIN-Farming online learning series, Roxanne's current role is to attract and support new farming talent. She contends that SPIN-Farming is uniquely suited to entrepreneurs and provides a career path for those who have a calling to farm. It is enticing first generation farmers who are keenly interested in matters of principle, but who understand that to have a significant positive impact, they have to function within the existing system, pushing their cause while paying their bills.

"For aspiring farmers, SPIN-Farming eliminates the 2 big barriers to entry – sizeable acreage and substantial startup capital. At the same time, its intensive relay growing techniques and precise revenue targeting formulas push yields to unprecedented levels and result in highly profitable income," Roxanne says. "While most other farming systems focus primarily if not exclusively on agricultural practices, SPIN-Farming emphasizes the business aspects and provides a financial and management framework for having the business drive the agriculture, rather than the other way around."

As SPIN-Farming becomes established and is practiced more and more widely, Roxanne says, it will create new farmland closer to metropolitan areas, which, in turn will produce environmental, economic and social benefits. "It offers a compelling value proposition."

"We can't explain GAIA, but we can explain commercial coolers."

The Beginning…

www.ingramcontent.com/pod-product-compliance
Lightning Source LLC
Chambersburg PA
CBHW052052190326
41519CB00002BA/195